Published by Fox College of Business

© **Bryan K. Law, 2012**

Law, Bryan K.

Shortcut Reasoning ®

Mathematics Series I – Tricks & Games

Includes index.
ISBN 978-0-9809409-9-2

1. Mathematics   2. Games

Produced in Canada

Bryan K. Law, BSc (Math), LLM

A graduate of the Chinese University of Hong Kong with major in mathematics and minor in statistics, Bryan got his master of laws degree from the University of Northumbria, specializing in commercial property law. He is now pursuing the doctor of laws degree from UNISA.

Bryan was a high school math teacher before beginning his career in the business world. After years of employment, entrepreneurship and continuing education, he knows exactly how a mathematical mind can assist one's academic results, career advancements and business ventures.

Besides the role as a professor and program coordinator of Fox College of Business, Bryan is also a certified instructor of OREA Real Estate College, responsible for instructing the real estate licensing courses in Ontario.

Bryan is the author of the best selling books *Feng Shui 123, Real Estate; Every One Can Afford It!; Basic Feng Shui Guide* and numerous articles in journals, newspapers and newsletters.

Table of Contents

Preface

Having a mathematical mind is the utmost important thing for everyone, as you need a logical mind and the ability to respond quickly in most of our daily activities.

Studying any math subject involves logical thinking and systematic analysis. Students should be patient in reading the text, understanding the concepts, doing the exercises and should repeat such routines when needed.

This book is not a text book; instead, it serves as an essential guide to understanding basic math techniques in a simplified and tightly focused manner. It provides fast and effective methods to readers who need concise material to assist them in basic computations.

This book was written with the main goal in mind – to improve your mathematical skills in an easy and enjoyable manner. Whether you are a student or a working adult who now wants to improve and polish your basic math skills, I hope that you will find what follows both effective and interesting.

The provided games help you internalize the skills with fun. I believe that the best state of mind for learning is to have fun and be relaxed. The games will help serve the purpose.

I would like to use this preface to thank my wife, Gladys, for her input and my daughter, Tania, for proofreading.

Bryan K. Law, Summer 2012

Introduction

What is mathematics?

The answer will be different; it depends on the level of education. The Random Webster's Unabridged Dictionary defines mathematics as "*the systematic treatment of magnitude, relationships between figures and forms, and relations between quantities expressed symbolically*".

The online encyclopedia, Wikipedia, gives mathematics the definition as "*the study of quantity, structure, space, and change. Mathematicians seek out patterns, formulate new conjectures, and establish truth by rigorous deduction from appropriately chosen axioms and definitions.*"

The impression of the general public may be that mathematics is a science subject dealing with the logic of quantity, shape and arrangement. The well-known scientist Albert Einstein, who is generally acknowledged as the smartest guy in the world once said: "*Pure mathematics is, in its way, the poetry of logical ideas.*"

We can see that logic or logical thinking plays a very important part in mathematics. However, the 'mathematics' that the majority of people apply daily is just arithmetic, another foundation of mathematics. Although daily arithmetic involves only simple calculations of the basic operation – addition, subtraction, multiplication and division, it is already 'complicated' enough to scare many people.

Carl Friedrich Gauss, one of the greatest mathematicians and the Titan of science, said: *"Mathematics is the queen of sciences and arithmetic is the queen of mathematics."* Although the arithmetic Gauss referred to was, in fact, the number theory, still you can tell how important arithmetic is. It is, therefore, crucial for us to be familiar with the basics of arithmetic.

Mathematics in Our Daily Life

Roger Bacon, philosopher and scientist of the 13th century, once stated: *"Neglect of mathematics works injury to all knowledge since he who is ignorant of it cannot know the other sciences or the things of the world. And what is worst, those who are thus ignorant are unable to perceive their own ignorance and so do not seek a remedy."*

This is a very true statement; as mathematics is the core of almost everything. In fact, we always apply mathematics in our daily life without noticing it. For example, we have to calculate the discount rates when we shop; to find out how much the change is when we pay the cashier in a store; to follow a recipe with different quantities; to finance or lease a car; to calculate the unit cost when buying groceries; and find out the directions and distances in navigating. All of these involve basic mathematical concepts or simple calculations.

Get Rid of the Calculator!

One of the reasons why people do not get good mathematics skills is that they rely too much on the calculator. Using a calculator is a fast way to get the answer of a lengthy calculation, but it may take a longer time in most of the simple calculations.

All people with good math skills can do simple multiplications and divisions without using a calculator. Actually, they can do them faster than those using a calculator; as it takes time to key in the numbers.

In fact, doing calculations without using a calculator helps you tune up your brain so that your brain cells remain active. Just try to calculate the questions below without using a calculator (answers are on the next page).

a) 12×12

b) 25×25

c) $100 \div 25$

d) $40 with a discount of 30% less

e) $4 \times 5 + 3 \times 7$

f)   75% of 400

Answers:

a)   144

b)   625

c)   4

d)   28

e)   41

f)   300

If you can do them all correctly within two minutes, it means your math skills are quite good already. If you do not get all four correct answers or it takes more than one minute to finish them; don't worry, you will be able to get all answers correctly within a minute after reading this book, by knowing the tricks and playing the games.

Part I – Basic Techniques

Get Familiar with Numbers

We all learned numbers since we were toddlers, but not all of us are familiar with the numbers. The reason why some of us can play with numbers better than the others is that we could visualize the numbers at a younger age and then keep such visualization for a lifetime.

Kindergartens use different methods and games to assist preschoolers in memorizing and visualizing numbers. Although most of the methods used are good, students stop using such methods or games once they think they are capable of applying for the numbers or once they enter elementary school. As a result, their skills deteriorate as they grow older.

In order to keep our sensitivity to numbers, we should stay familiar with numbers by visualizing them with the assistance of simple games and calculating simple questions without using a calculator.

Numbers 1 to 10

To make mathematics an easy subject, it should be added to one's activities as early as possible. Arithmetic is actually a natural thing that can be learnt without going to school. Since we use the decimal system in daily life, it is important to know how to count from 1 to 10, from 1 to 100, and by multiples of 10 and so on. Therefore, toddlers and preschoolers should get familiar with numbers by counting from 1 to 10 and then backwards from 10 to 1 through activities.

The best way to get familiar with number sense is to use a deck of playing cards to play with the numbers. We will only use the cards from ace to 10, and ace will be treated as 1. Once they can visualize the numbers with the cards, they can play with the numbers easily when they get older.

After they have known the numbers well, they can play some card games in Part II of this book.

Exercise I (Answers are on the next page)

Try to calculate the following questions without using a pen and calculator (you can use a pen to write down the answers, but not to calculate):

a)   5 + 7

b)   6 + 9

c)   13 + 8

d)   26 + 17

e)   34 + 49

f)   128 + 336

g)   11 – 7

h)   13 – 6

i)   41 – 25

j)   81 – 52

k)   111 – 55

l)   204 – 125

Answers to Exercise I

a)   12

b)   15

c)   21

d)   43

e)   83

f)   464

g)   4

h)   7

i)   16

j)   29

k)   56

l)   79

Did you get them all right in two minutes? If not, don't worry, you should be able to do it after practicing the exercises and playing the games in this book in a few weeks time or sooner.

Multiplication Table

It is crucial to memorize the multiplication table (from 1 to 10) as you will be applying it everyday. Unfortunately, there is no shortcut to memorize it; you have to memorize them one by one.

Although we say the multiplication table is from 1 to 10, you only need to memorize it from 2 to 9 as the multiples of 1 and 10 are obvious. Multiplying 1 to any number will result in the product being the number itself, multiplying 10 to any number will result in a product with a zero added to the right.

Examples:
$$1 \times 11 = 11$$
$$1 \times 381 = 381$$
$$10 \times 14 = 140$$
$$10 \times 282 = 2820$$

The word table (Figures 1 & 2) is for the purpose of saying them out loud word-by-word so that you can memorize it easier. For example, you should repeatedly say the words exactly as shown in the table – "Two-One Two, Two-Two Four, Two-Three Six, Two-Four Eight, …" until you can memorize all of them.

Once you are familiar with the word table; you can use the number table (Figure 3) to strengthen your memory. The number table is to visualize the multiplication table so that you will have a faster reaction in mind. Numbers are more effective than words when you have to call them out from your memory. You can also use the number table to read the numbers aloud as if it is the word table: "2-1, 2; 2-2, 4; 2-3, 6; 2-4, 8; …"

When can you stop practicing? You should keep on practicing until you can remember the answers in less than one second for all single digits multiplications.

Multiplication Table (Words) - I

Two-One Two	Three-One Three	Four-One Four	Five-One Five
Two-Two Four	Three-Two Six	Four-Two Eight	Five-Two Ten
Two-Three Six	Three-Three Nine	Four-Three Twelve	Five-Three Fifteen
Two-Four Eight	Three-Four Twelve	Four-Four Sixteen	Five-Four Twenty
Two-Five Ten	Three-Five Fifteen	Four-Five Twenty	Five-Five Twenty Five
Two-Six Twelve	Three-Six Eighteen	Four-Six Twenty Four	Five-Six Thirty
Two-Seven Fourteen	Three-Seven Twenty One	Four-Seven Twenty Eight	Five-Seven Thirty Five
Two-Eight Sixteen	Three-Eight Twenty Four	Four-Eight Thirty Two	Five-Eight Forty
Two-Nine Eighteen	Three-Nine Twenty Seven	Four-Nine Thirty Six	Five-Nine Forty Five
Two-Ten Twenty	Three-Ten Thirty	Four-Ten Forty	Five-Ten Fifty

Figure 1

Multiplication Table (Words) - II

Six-One Six	Seven-One Seven	Eight-One Eight	Nine-One Nine
Six-Two Twelve	Seven-Two Fourteen	Eight-Two Sixteen	Nine-Two Eighteen
Six-Three Eighteen	Seven-Three Twenty One	Eight-Three Twenty Four	Nine-Three Twenty Seven
Six-Four Twenty Four	Seven-Four Twenty Eight	Eight-Four Thirty Two	Nine-Four Thirty Six
Six-Five Thirty	Seven-Five Thirty Five	Eight-Five Forty	Nine-Five Forty Five
Six-Six Thirty Six	Seven-Six Forty Two	Eight-Six Forty Eight	Nine-Six Fifty Four
Six-Seven Forty Two	Seven-Seven Forty Nine	Eight-Seven Fifty Six	Nine-Seven Sixty Three
Six-Eight Forty Eight	Seven-Eight Fifty Six	Eight-Eight Sixty Four	Nine-Eight Seventy Two
Six-Nine Fifty Four	Seven-Nine Sixty Three	Eight-Nine Seventy Two	Nine-Nine Eighty One
Six-Ten Sixty	Seven-Ten Seventy	Eight-Ten Eighty	Nine-Ten Ninety

Figure 2

Multiplication Table (Numbers)

2x1 = 2	3x1 = 3	4x1 = 4	5x1 = 5
2x2 = 4	3x2 = 6	4x2 = 8	5x2 = 10
2x3 = 6	3x3 = 9	4x3 = 12	5x3 = 15
2x4 = 8	3x4 = 12	4x4 = 16	5x4 = 20
2x5 = 10	3x5 = 15	4x5 = 20	5x5 = 25
2x6 = 12	3x6 = 18	4x6 = 24	5x6 = 30
2x7 = 14	3x7 = 21	4x7 = 28	5x7 = 35
2x8 = 16	3x8 = 24	4x8 = 32	5x8 = 40
2x9 = 18	3x9 = 27	4x9 = 36	5x9 = 45
2x10 = 20	3x10 = 30	4x10 = 40	5x10 = 50

6x1 = 6	7x1 = 7	8x1 = 8	9x1 = 9
6x2 = 12	7x2 = 14	8x2 = 16	9x2 = 18
6x3 = 18	7x3 = 21	8x3 = 24	9x3 = 27
6x4 = 24	7x4 = 28	8x4 = 32	9x4 = 36
6x5 = 30	7x5 = 35	8x5 = 40	9x5 = 45
6x6 = 36	7x6 = 42	8x6 = 48	9x6 = 54
6x7 = 42	7x7 = 49	8x7 = 56	9x7 = 63
6x8 = 48	7x8 = 56	8x8 = 64	9x8 = 72
6x9 = 54	7x9 = 63	8x9 = 72	9x9 = 81
6x10 = 60	7x10 = 70	8x10 = 80	9x10 = 90

Figure 3

Exercise II (Answers are on the next page)

Try to calculate the following questions without using a pen and calculator (you can use a pen to write down the answers, but not to calculate):

a)   4 × 7

b)   6 × 8

c)   8 × 3

d)   7 × 6

e)   5 × 5

f)   8 × 2

g)   6 × 9

h)   3 × 6

i)   9 × 5

j)   11 × 5

k)   7 × 12

l)   4 × 25

Answers to Exercise II

a) 28

b) 48

c) 24

d) 42

e) 25

f) 16

g) 54

h) 18

i) 45

j) 55

k) 84

l) 100

You should have gotten all answers correctly in less than one minute. If not, please practice the multiplication table again until you are very familiar with them.

Quarters

It is important to know what a quarter means. Yes, it is one fourth of a unit, but how much is one fourth?

A quarter dollar is 25 cents, a quarter hour is 15 minutes, and a quarter dozen is 3. It may confuse the kids at the very beginning; especially the first time they learn how to tell time and count money. The important message of a quarter is that it is 25% of a unit.

One day I was buying a cup of coffee which cost $1.78 including tax. Since I had too many coins in my pocket, I handed the cashier $2.03; the lady looked very puzzled and gave the 3 cents back to me. I did not take the 3 cents and asked her to punch in $2.03 and then see the change on the cash register. She could not figure out what the point was but just followed my request. She punched in $2.03 and found out from the cash register that the change was exactly 25¢. She smiled and handed me a quarter as the change.

Such kind of situations happens all the time. When a cashier gives back one's pennies or looks puzzled; you know that cashier is 'not so smart'. The cashier has to rely on the calculator (cash register) to help rationalize what the customer wants. On the other hand, through quick mental calculations, the customer hands the cashier extra pennies in order to receive fewer coins back. A better mathematically trained person solves such kind of problems much more quickly.

The ability to visualize the concept of a quarter is useful in daily applications. You should get familiar with the concept of a quarter as soon as possible.

Exercise III (Answers are on the next page)

Try to calculate the following questions without using a pen and calculator (you can use a pen to write down the answers, but not to calculate):

a)   How many quarters do you have when you change a $10 bill into quarters.

b)   How many minutes are there in one and quarter hours (one hour has 60 minutes)?

c)   A pen cost you $1.77 and the change was $0.25. How much did you pay?

d)   You have 19 quarter coins, how much money do you have?

e)   You have $6 and want to change it to all quarter coins. How many quarter coins may you get?

f)   The parking fee is $0.25 per 15 minutes. How much will you have to pay if you park your car there for 10 hours?

Answers to Exercise III

a)   40 quarters

b)   75 minutes

c) $2.02

d) $4.75

e)   24

f) $10

Again, you should be able to get all the correct answers in one minute.

Short Cut Summary

➢ Playing with a deck of cards is a good way to visualize numbers.

➢ Mastering the multiplication table is essential in doing multiplication and division calculations.

➢ The concept of a quarter is important in daily life.

1 quarter = $\dfrac{1}{4}$ (i.e. ÷4) or 25%.

Example: If you need to pay $1.78, you can pay the cashier $2.03 to receive a change of a quarter. You have fewer coins to carry and look smarter.

Simple Tricks

Since we have to deal with simple calculations everyday, it is important to know some tricks to find the answers quicker. The faster you can get the answers, the better trained your brain becomes, and the smarter you will be in problem solving.

Addition and Subtraction

All basic calculations involve addition and subtraction. Since we use the decimal system, all operations involve an addition to 10 and subtraction from 10. It is therefore important to be familiar with the pairs within numbers 1 to 9 that pair up to make a sum of 10. The more familiar you are with the pairs, the faster you can calculate. For example,

1 and 9 is a pair as $1 + 9 = 10$

2 and 8 is a pair as $2 + 8 = 10$

3 and 7 is a pair as $3 + 7 = 10$

4 and 6 is a pair as $4 + 6 = 10$

5 and 5 is a pair as $5 + 5 = 10$

Once you are familiar with them, you can do the addition and subtraction much faster and will be able to move on to the next level – working with 100.

One of the most common applications for addition to 100 and subtraction from 100 is doing percentage calculations. As 1 is a whole, which also means 100%; familiarity with how to add up to or subtract from 100 helps you solve a lot of simple questions.

It is important to know which two numbers pair up together to make up a sum of 9 and 10, for this knowledge makes it easier and faster to calculate two digit sums to 100 and deduction from 100.

We can tell right away if two numbers will add up to 100 or not by seeing whether the tens digits add up to 9 and the ones digits add up to 10. Similarly, when we subtract a two-digit number from 100, it will be very easy to tell the answer if you can identify the tens digit pair and ones digit pair immediately.

For example, we can tell $37 + 63 = 100$ easily because we see that in the tens digit place, $3 + 6 = 9$ and in the ones digit place, $7 + 3 = 10$. For $100 - 42$, we know that the answer is 58 right away as $9 - 4 = 5$ and $10 - 2 = 8$.

Exercise IV (Answers are on the next page)

Try to calculate the following questions without using a pen and calculator (you can use a pen to write down the answers, but not to calculate):

a) $100 - 31 = ?$

b) $100 - 24 = ?$

c) $100 - 47 = ?$

d) $100 - 52 = ?$

e) $100 - 78 = ?$

f) $100 - 69 = ?$

g) $14 + ? = 100$

h) $83 + ? = 100$

i) $91 + ? = 100$

j) $33 + ? = 100$

k) $44 + ? = 100$

l) $65 + ? = 100$

m) $72 + ? = 100$

Answers to Exercise IV

a)   69

b)   76

c)   53

d)   48

e)   22

f)   31

g)   86

h)   17

i)   9

j)   67

k)   56

l)   35

m)   28

Again, you should be able to get all the correct answers in one minute.

Multiplication

Once you are familiar with the multiplication table (1 to 10), you should have no problem handling simple multiplications. However, your multiplication skills can be fine tuned and raised to another level by knowing some more techniques.

You may have found out from the multiplication table that the number 9, when multiplying by a number between 2 and 9, has a special pattern. The product of any single digit number and 9 will make a two digit number when the two digits are added together, the sum is 9. For example, $2 \times 9 = 18$ and the tens digit '1' plus the ones digit '8' will make a 9. Similarly, we have:

$$3 \times 9 = 27 \ ('2' + '7' = '9')$$
$$4 \times 9 = 36 \ ('3' + '6' = '9')$$
$$5 \times 9 = 45 \ ('4' + '5' = '9')$$
$$6 \times 9 = 54 \ ('5' + '4' = '9')$$
$$7 \times 9 = 63 \ ('6' + '3' = '9')$$
$$8 \times 9 = 72 \ ('7' + '2' = '9')$$
$$9 \times 9 = 81 \ ('8' + '1' = '9')$$

By knowing this, you should now have more confidence when calculating the single digit multiplications of 9 as it is very easy to double check your answers.

The number 5 also has its characteristic in multiplication. When you have to calculate the square of a 2-digit number ending in 5 (the square of a number is the product of the number multiplied to itself), you can simply multiply the first digit by the next consecutive number and put 25 as the last two digits. For example, 25 × 25 = (2×3)25 = 625; 35 × 35 = (3×4)25 = 1225 and 45 × 45 = (4×5)25 = 2025.

Although the metric system is widely used now, there are still some non-metric matters we have to deal with. Counting by dozens is one of them. By memorizing the table below (Figure 4), you can count and calculate items by the dozen more efficiently.

Getting familiar with the multiple of 12 can help you in calculating inches and feet, or other units with 12 as multiples, more effectively.

2 × 12	=	24
3 × 12	=	36
4 × 12	=	48
5 × 12	=	60
6 × 12	=	72
7 × 12	=	84
8 × 12	=	96
9 × 12	=	108
10 × 12	=	120
11 × 12	=	132
12 × 12	=	144
13 × 12	=	156
14 × 12	=	168
15 × 12	=	180
16 × 12	=	192
17 × 12	=	204
18 × 12	=	216
19 × 12	=	228
20 × 12	=	240

Figure 4

Exercise V (Answers are on the next page)

Try to calculate the following questions without using a pen and calculator (you can use a pen to write down the answers, but not to calculate):

a) 3×9

b) 6×9

c) 9×8

d) 85×85

e) 55×55

f) 95×95

g) 3×12

h) 12×5

i) 12×12

Answers to Exercise V

a) 27

b) 54

c) 72

d) 7225

e) 3025

f) 9025

g) 36

h) 60

i) 144

Division

Knowledge of the multiplication table is essential; likewise, we should also be very familiar with the division from 1 to 10. There is no set division table since a number is not divisible by any number. However, there are some rules can help us.

1 – Every number is divisible by 1 and the quotient (answer) is that number itself.

2 – Every even number is divisible by 2.

3 – A number will be divisible by 3 if the sum of all its digits is a multiple of 3.

For example, the number 35,431,452 is divisible by 3 as the sum of all its digits is a multiple of 3. If we add all the digits together, $3 + 5 + 4 + 3 + 1 + 4 + 5 + 2 = 27$ and 27 is divisible by 3. If you do not know whether 27 is divisible by 3 or not, just add 2 to 7 and you will get 9. Of course, you should know 3, 6, 9 are all divisible by 3.

In fact, two-digit numbers and three-digit numbers are the numbers we often come across. For example, you may dine out with two friends and when the bills come, you have to calculate the total cost and divide it by 3. For example, the total is $172.61 and you round it up to $173. However, 173 is not divisible by 3 as $(1 + 7 + 3)$ is 11. As a result, you may round it to $174 so that it is divisible by 3.

4 – A number will be divisible by 4 if the number formed by its last two digits is divisible by 4. For example, the

number 4,349,898,548 is divisible by 4 as the number formed by its last two digits is 48 and 48 is divisible by 4.

(Note: if the last 2 digits are 00, the number is divisible by 4 as it is a multiple of 100 and 100 is divisible by 4).

The numbers 4 and 25 are a pair. When the number 100 is divided by 4, the result is 25. This is why we have quarters – 25¢ is a ¼ dollar. The same division rule applies to 25; that is, if the last two digits of a number can be divided by 25, the number is divisible by 25. Clearly, we have only 4 possible scenarios – the two digits are 00, 25, 50 and 75.

5 – A number will be divisible by 5 if it ends with 0 or 5 (the last digit is 0 or 5). The numbers 2 and 5 are a pair. When you divide a number by 5, it is the same as multiplying it by 0.2.

For example, if you want to calculate 234,112 ÷ 5, it is easy to multiply the number by 2 and move the decimal place to the left (multiply by 0.2). You will get 46,822.4 right away without difficulty. Another example 12,340 ÷ 5 = 2,468 (i.e. 2,468.0, the .0 is omitted.)

6 – A number will be divisible by 6 if it is an even number and is divisible by 3 (see the trick above).

7 – A number will be divisible by 7 if the difference between the number of its last three digits and the number of all the preceding digits can be divided by 7.

Examples:

a) 16,002 is divisible by 7 as 16 – 2 is 14, and 14 is divisible by 7.

b) 3,066 is divisible by 7 as 66 – 3 is 63; and 63 is divisible by 7.

c) 100,107 is divisible by 7 as 107 – 100 is 7, and 7 is divisible by 7.

d) 560,511 is divisible by 7 as 560 – 511 is 49; and 49 is divisible by 7.

Note: If the difference is 0 (i.e.; the number is a 6-digit one with the first 3 digits the same as the last 3 digits), the number will be divisible by 7.

8 – A number will be divisible by 8 if the number formed by its last three digits can be divided by 8.

For example, the number 4,349,898,848 is divisible by 8 as the number formed by its last three digits is 848 and 848 is divisible by 8 (Note: if the last 3 digits are 000, the number is divisible by 8 as it is a multiple of 1000 and 1000 is divisible by 8).

The numbers 8 and 125 are a pair. When the number 1,000 is divided by 8, the result is 125. It is useful to remember 1/8 is 0.125, 2/8 is 0.25, 3/8 is 0.375, 4/8 is 0.5, 5/8 is 0.625, 6/8 is 0.75, and 7/8 is 0.875.

9 – A number will be divisible by 9 if the sum of all the digits is a multiple of 9. For example, the number 53,438,652 is divisible by 9 as the sum of all its digits 5 + 3 + 4 + 3 + 8 + 6 + 5 + 2 = 36 and 36 is divisible by 9.

10 – Every number ending with the digit 0 is divisible by 10.

Exercise VI (Answers are on the next page)

Try to calculate the following questions without using a pen and calculator (you can use a pen to write down the answers, but not to calculate):

a)   7 × 12

b)   12 × 8

c)   11 × 11

d)   9 × 12

e)   12 × 11

f)   12 × 12

g)   Is 3,344,364 divisible by 3?

h)   Is 3,344,364 divisible by 6?

i)   Is 3,344,364 divisible by 9?

j)   Is 3,344,304 divisible by 4?

k)   Is 3,344,104 divisible by 8?

l)   Is 344,351 divisible by 7?

Answers to Exercise VI

a)   84

b)   96

c)   121

d)   108

e)   132

f)   144

g)   Yes. 3 + 3 + 4 + 4 +3 +6 +4 = 27, and 27 is divisible by 3.

h)   Yes. Since it is an even number and divisible by 3.

i)   Yes. 3 + 3 + 4 + 4 +3 +6 +4 = 27, and 27 is divisible by 9.

j)   Yes. The last two digits are 04 and 4 is divisible by 4.

k)   Yes. The last 3 digits are 104, and 104 is divisible by 8.

l)   Yes. The last 3 digits are 351 and the number preceding 351 is 344; the difference is 7, therefore is divisible by 7.

Short Cut Summary

➢ Familiar with the numbers to make up 10 can speed up basic addition and subtraction calculations.

- Examples:
- a)   2 numbers: 1 + 9, 2 + 8, 3 + 7, 4 + 6, 5 + 5
- b)   3 numbers: 1+ 4 + 5, 2 + 4 + 4, 6 + 3 + 1, 5 + 3 + 2

➢ Familiar with the pair to make up 9 and 10 can speed up additions and subtractions related to 100.

- Examples:
- a)   To make up 9: 1 + 8, 2 + 7, 3 + 6, 4 + 5
- b)   To make up 10: 1 + 9, 2 + 8, 3 + 7, 4 + 6, 5 + 5

➢ Remember 1 = 100%. A good mastery of adding a pair to 100 and subtraction a number from 100 is essential for daily application of mathematics.

➢ It is helpful to memorize multiplication for 12 (i.e. dozens).

➢ A few tricks and multiplication and division:

o   Numbers with 0 as its last digit are divisible by 10.

o   All even numbers are divisible by 2.

o   A number will be divisible by 3 if the sum of all its digits is a number divisible by 3.

- E.g. 35,431,452 – the sum of all digits is 27, a multiple of 3. Therefore, 35,431,452 is divisible by 3.

- A number will be divisible by 4 if the number formed by its last 2 digits is divisible by 4. Example: 4,349,898,548 is divisible by 4 since the number formed by its last 2 digits is 48 and 48 is divisible by 4. Note: If its last 2 digits are 00, it is also divisible by 4.

- A number will be divisible by 5 if the ending digit is either 0 or 5. To divide a number by 5, it may be easier to do it by multiplying by 0.2. E.g. 234,112 ÷ 5 = 26,822.4 (multiply by 2, and move 1 decimal place to the left.)

- A number will be divisible by 6 if it is an even number and sum of digits is divisible by 3 (since 6 = 2 × 3). Example: 35,431,452 is divisible by 6.

- A number will be divisible by 7 if the difference between its last 3 digits and all the preceding digits is divisible by 7. E.g. 100,107 is divisible by 7 since the difference of 107 and 100 is 7 (107,100 is also divisible by 7). Note: if the difference is 0, the number will also be divisible by 7.

- A number will be divisible by 8 if the number formed by its last 3 digits is divisible by 8 (similar to 4).

- A number will be divisible by 9 if the sum of all its digits is a number that is divisible by 9 (similar to 3).

- The square of 2-digit numbers ending with 5 is the number formed by multiplying the first digit and its next consecutive number and put 25 as the last two digits.
  e.g.

  - $15 \times 15 = (1 \times 2)\ 25 = 225$
  - $25 \times 25 = (2 \times 3)\ 25 = 625$
  - $35 \times 35 = (3 \times 4)\ 25 = 1225$
  - $45 \times 45 = (4 \times 5)\ 25 = 2025$

How to Switch Numbers

Mathematics always involves equations and we have to solve them. All equations have an equal sign in it and solving the equations will require moving the numbers or variables from one side of the equal sign to the other side. For example:

$$x - 4 = 0$$

$$\Rightarrow \quad x = 4$$

(We moved the number 4 from left hand side to right hand side of the equal sign, and the –4 became a positive 4)

There are rules that apply to such kind of calculations so that you can compute the equations more efficiently.

Rule #1

Addition becomes subtraction and subtraction becomes addition when switching a number or variable from one side of the equal sign to the other side. That is, the + and – signs change when you switch the numbers.

Examples:

a) $$x - 3 = 5$$

$$\Rightarrow \quad x = 5 + 3$$

(– 3 becomes + 3 when it moves from left to right)

Therefore, x = 8.

b) $4 - y = 0$

$\Rightarrow$ $4 = 0 + y$

(– y becomes + y when it moves from left to right)

Therefore, y = 4.

c) $x - 3y = 5y - 2x$

$\Rightarrow$ $x + 2x = 5y + 3y$

$\Rightarrow$ $3x = 8y$

(– 2x becomes + 2x when it moves from right to left,
– 3y becomes + 3y when it moves from left to right)

Rule #2

Multiplication becomes division and division becomes multiplication when switching a number or variable from one side of the equal sign to the other side. That is, the × and ÷ signs change when you switch the numbers.

Examples:

a) $3x = 15$

$\Rightarrow \qquad x = 15 \div 3$

$x = 5$

(3x means 3 multiplies x or x multiplies 3, the 3 becomes ÷ 3 when it moves from left to right)

b) $\qquad y \div 2 = 7$

$\Rightarrow \qquad y = 7 \times 2$

$y = 14$

(÷ 2 becomes × 2 when it moves from left to right)

Rule #3

When fractions are involved, the best method is to use cross multiplications: multiply the numerator on the left hand side by the denominator on the right hand side, and multiply the numerator on the right hand side by the denominator on the left hand side.

Examples:

a) $\qquad \dfrac{3x}{7} = \dfrac{2}{5}$

$$\frac{3x}{7} \diagdown\!\!\!\!\diagup \frac{2}{5}$$ (multiply 3x by 5 and 2 by 7)

$$\Rightarrow \quad 15x = 14$$

Therefore, x = $\dfrac{14}{15}$

b) $\qquad \dfrac{4}{5x} = \dfrac{2}{3}$

$$\frac{4}{5x} \diagdown\!\!\!\!\diagup \frac{2}{3}$$ (multiply 4 by 3 and 2 by 5x)

$$\Rightarrow \quad 12 = 10x$$

$$\Rightarrow \quad x = \frac{12}{10}$$

Therefore x = $\dfrac{5}{6}$

c) $\qquad \dfrac{1}{x} = \dfrac{5}{4}$

$$\frac{1}{x} \diagdown\!\!\!\!\diagup \frac{5}{4}$$ (multiply 1 by 4 and 5 by x)

$$\Rightarrow \quad 4 = 5x$$

$$\Rightarrow \quad x = \frac{4}{5}$$

Note: You can get the answer much quicker by reversing the fractions (flipping them upside down):

$$\frac{1}{x} = \frac{5}{4} \qquad \Rightarrow \qquad \frac{x}{1} = \frac{4}{5}$$

$$\Rightarrow \qquad x = \frac{4}{5}$$

Exercise VII (Answers are on the next page)

Find x in all questions:

a)   x − 9 = 15

b)   3 − x = 2

c)   5 + 2x = x + 11

d)   − x = 2x − 9

e)   3x − 4 = 4 − x

f)   3x − 2 = 7

g) $\dfrac{7}{4x} = 2$

h) $\dfrac{4}{x+3} = 1$

i) $\dfrac{2x+4}{x-3} = 12$

j) $3x - 1 = \dfrac{2x+3}{3}$

k) $\dfrac{3}{(4x-1)} = \dfrac{1}{(x+2)}$

Answers to Exercise VII

a) $x = 15 + 9$
         $x = 24$

b) $3 - 2 = x$
         $x = 1$

c) $2x - x = 11 - 5$
         $x = 6$

d) $9 = 2x + x$
         $9 = 3x$
         $x = 3$

e) $3x + x = 4 + 4$
         $4x = 8$
         $x = 2$

f) $3x = 9$
         $x = 3$

g) $7 = 8x$
         $x = \dfrac{7}{8}$

h) $4 = x + 3$
         $x = 4 - 3$
         $x = 1$

i) $2x + 4 = 12 (x - 3)$
         $2x + 4 = 12x - 36$
         $40 = 10x$
         $x = 4$

j) $3 (3x - 1) = 2x + 3$
         $9x - 3 = 2x + 3$
         $7x = 6$
         $x = \dfrac{6}{7}$

k) $3(x + 2) = 4x - 1$
         $3x + 6 = 4x - 1$
         $7 = x$

Short Cut Summary

➢ Simple rules for switching numbers to the other side of the equation.

- o Rule #1: Adding becomes subtracting, subtracting becomes adding (positive becomes negative, negative becomes positive).

  - ▪ Examples:

    a) $x - 3 = 0$
    $x = 3$

    b) $8 - y = 7$
    $8 - 7 = y$
    $y = 1$

- o Rule # 2: Multiplying becomes dividing, dividing becomes multiplying.

  - ▪ Examples:

    a) $3x = 8$
    $x = \dfrac{8}{3}$

    b) $\dfrac{16}{5x} = 2$
    $16 = 10x$
    $x = \dfrac{16}{10}$
    $x = \dfrac{8}{5}$

- Rule # 3: For a fraction, use the cross multiplication. That is, multiplying the denominator to the numerator of the other side.

  - Example:
  $$\frac{3x}{7} = \frac{2}{5}$$
  $$15x = 14$$
  $$x = \frac{14}{15}$$

- Rule # 4: You can also reverse the fractions (flipping them) if it can give you the answer faster.

  - Example:
  $$\frac{1}{x} = \frac{2}{9}$$
  $$x = \frac{9}{2}$$

The Percentage

The percentage is one of the most frequently used mathematical tools in our daily life. By mastering the technique in calculating the percentage, you can improve your computation skills in almost every subject.

Discount

One of the most common applications of percentage is to calculate discount offered by stores and merchants. For example, a pen is selling at $90 with 30% discount; how much is the discounted price?

Traditionally, we will calculate how much is that 30% and then deduct that 30% value from the original price to get the discounted price. The steps are:

$90 × 30% = $27 (the discount you enjoy)

Discounted price will then be the original price minus the discount:

Discounted price = $90 – $27

= $63

A faster way will be to apply the discount 'directly' to the listing price by knowing the net percentage (100% – the discount rate) and multiply it to the original price. Since doing subtraction from 100(%) is much easier than the calculations above, the calculation below is much faster.

$$100\% - 30\% = 70\% \text{ (the net percentage)}$$

$$\$90 \times 70\% = \$63 \text{ (the discounted price)}$$

It is therefore important to be skillful in adding up a pair to 100 and subtracting a number from 100. You should repeat the topics of addition and subtraction in **Simple Tricks** section.

In some cases, you may want to just estimate the discounted price when the original price is not a 'perfect' number. For example, the prices $9.95, $19.90 and $199 may be rounded to $10, $20 and $200 respectively when you just want to estimate how much the discounted prices will be, so that you can tell if it is a good bargain or not.

Exercise VIII (Answers are on the next page)

Try to calculate the following questions without using a pen and calculator (you can use a pen to write down the answers, but not to calculate):

a)   What is the discounted price of a pen, when the listing price is $100 and the discount is 35%?

b)   What is the discounted price of a bag, when the listing price is $700 and the discount is 40%?

c)   What is the discounted price of a book, when the listing price is $45.50 and the discount is 90%?

d)   What is the discounted price of a computer, when the listing price is $2,426.50 and the discount is 50%?

e)   What is the estimated discounted price (to the nearest dollar) of a computer, when the listing price is $1,999.90 and the discount is 10%?

f)   What is the estimated discounted price (to the nearest dollar) of a shirt, when the listing price is $99.90 and the discount is 35%?

g)   What is the estimated discounted price (to the nearest dollar) of a bag, when the listing price is $19.90 and the discount is 20%?

Answers to Exercise VIII

a) $65 ($100 × 65%)

b) $420 ($700 × 60%)

c) $4.55 ($45.50 × 10%)

d) $1,213.25 (50% of $2,426.50 is 1/2 × 2,426.50)

e) $1,800 (round up $1,999.90 to $2,000 and multiply it by 90%).

f) $65 (round up $99.90 to $100 and multiply it by 65%).

g) $16 (round up $19.90 to $20 and multiply it by 80%).

Percentage, decimal and Fraction

It is important to understand the relationships among percentage, decimal and fraction. If you can familiar the relationship among them, you may be able to do some multiplications and divisions more efficiently.

Percentage is to divide 1 by 100, so that 1% is 1/100, which is the same as 0.01. You should memorize the following teams:

a) 0.05 = 5% = 1/20
b) 0.1 = 10% = 1/10
c) 0.2 = 20% = 1/5
d) 0.25 = 25% = 1/4
e) 0.5 = 50% = 1/2

a) Multiplying a number by 5% (0.05) is the same as dividing it by 20. For example:

i) $200 \times 5\% = 200 \times 0.05$
$= 200 \div 20$
$= 10$

Dividing a number by 5% (0.05) is the same as multiplying it by 20. For example:

ii) $11 \div 5\% = 11 \div 0.05$
$= 11 \times 20$
$= 220$

b) Multiplying a number by 10% (0.1) is the same as dividing it by 10, or simply by dropping the last zero or moving the decimal point to the left by one digit. For example:

    iii) $32{,}400 \times 10\% = 32{,}400 \times 0.1$
                           $= 32{,}400 \div 10$
                           $= 3{,}240$ (dropping the last zero)

    iv) $233.25 \times 10\% = 233.25 \times 0.1$
                           $= 233.25 \div 10$
                           $= 23.325$ (moving the decimal point to the left by one digit)

Dividing a number by 10% (0.1) is the same as multiplying it by 10, or simply by adding a zero to it or moving the decimal point to the right by one digit. For example:

    v) $32{,}400 \div 10\% = 32{,}400 \div 0.1$
                         $= 32{,}400 \times 10$
                         $= 324{,}000$ (adding a zero to it)

    vi) $233.25 \div 10\% = 233.25 \div 0.1$
                           $= 233.25 \times 10$
                           $= 2332.5$ (moving the decimal point to the right by one digit)

c)   Multiplying a number by 20% (0.2) is the same as dividing it by 5. For example:

    vii) $255 \times 20\% = 255 \times 0.2$
                        $= 255 \div 5$
                        $= 51$

viii) $21 \div 20\%$ $= 21 \div 0.2$
                              $= 21 \times 5$
                              $= 105$

Dividing a number by 20% is the same as multiplying it by 5. For example:

ix) $40 \div 20\%$ $= 40 \div 0.2$
                            $= 40 \times 5$
                            $= 200$

x) $1 \div 20\%$ $= 1 \div 0.2$
                          $= 1 \times 5$
                          $= 5$

d)   Multiplying a number by 25% (0.25) is the same as dividing it by 4. For example

xi) $24 \times 25\%$ $= 24 \times 0.25$
                              $= 24 \div 4$
                              $= 6$

xii) $80 \times 25\%$ $= 80 \times 0.25$
                                $= 80 \div 4$
                                $= 20$

Dividing a number by 25% (0.25) is the same as multiplying it by 4. For example:

xiii) $3 \div 25\%$ $= 3 \div 0.25$
                            $= 3 \times 4$
                            $= 12$

xiv) $11 \div 25\% = 11 \div 0.25$
$= 11 \times 4$
$= 44$

e) Multiplying a number by 50% is the same as dividing it by 2 (such as half price sale);

xv) $2{,}440 \times 50\% = 2{,}440 \times 0.5$
$= 2{,}440 \div 2$
$= 1{,}220$

xvi) $428 \times 50\% = 428 \times 0.5$
$= 428 \div 2$
$= 214$

Other pairs to be memorized are:

f) $0.04 \quad = 4\% \quad = 1/25$
g) $0.125 = 12.5\% = 1/8$

Exercise IX (Answers are on the next page)

Try to calculate the following questions without using a pen and calculator (you can use a pen to write down the answers, but not to calculate):

a)   20% of 55

b)   50% of 20

c)   25% of 444

d)   12.5% of 8,888

e)   10% of 243.3

f)   40 ÷ 0.25

g)   11,223 ÷ 0.5

h)   411 ÷ 0.2

i)   322.4 ÷ 0.1

j)   2 ÷ 5%

Answers to Exercise IX

a)　　11　　(20% of 55 = 55 ÷ 5)

b)　　10　　(50% of 20 = 20 ÷ 2)

c)　　111　　(25% of 444 = 444 ÷ 4)

d)　　1,111　(12.5% of 8,888 is the same as 1/8 of 8,888)

e)　　24.33　(move the decimal point of 243.3 to left by one point)

f)　　160　　(40 ÷ 0.25 = 40 × 4)

g)　　22,446 (11,223 ÷ 0.5 = 11,223 × 2)

h)　　2055　(411 ÷ 0.2 = 411 x5)

i)　　3224　(move the decimal point of 322.4 to right by one point)

j)　　40　　(2 ÷ 5% = 2 × 20)

Short Cut Summary

Percentage application is common in daily life.

➢ An easier way to calculate discounted price is to subtract the discount percentage from 100% and multiply it by the original price.

- o E.g. $150 with 40% off.
  Discounted Price = $150 × 60% = $90

➢ Simple tricks for multiplying and dividing percentages are:

- o 5% (i.e. 0.05) = 1/20.
  Multiplying by 5% = dividing by 20.
  Dividing by 5% = multiplying by 20.

- o 10% (i.e. 0.1) = 1/10.
  Multiplying by 10% = dividing by 10. You move the decimal point to the left by one digit.
  Dividing by 10% = multiplying by 10. You move the decimal point to the right by one digit.

- o 12.5% (i.e. 0.125) = 1/8.
  Multiplying by 12.5% = dividing by 8.
  Dividing by 12.5% = multiplying by 8.

- o 20% (i.e. 0.2) = 1/5.
  Multiplying by 20% = dividing by 5.
  Dividing by 20% = multiplying by 5.

- 25% (i.e. 0.25) = 1/4.
  Multiplying by 25% = dividing by 4.
  Dividing by 25% = multiplying by 4.

- 50% (i.e. 0.5) = 1/2.
  Multiplying by 50% = dividing by 2.
  Dividing by 50% = multiplying by 2.

Change in Percentage

We are in an information era and everyday we hear the news 'Dow Jones industrial average down 2.8 percent' or 'Consumer Price Index rises 5.4%' or 'The fuel price climbed 5.5 percent over night'. These are the news talking about changes and differences in prices or index.

It is an easy task for us to understand the data when it is presented to us on the newspaper – 'Dow Jones industrial average down 2.8 percent'. We know that the Dow Jones goes down by 2.8 percent. 'Consumer Price Index rises 5.4%' means the CPI increases by 5.4%. However, when we are given only the raw data (only the numbers); it may be a challenge to tell how much is the increase or decrease in percentage if we do not know the basics of change in percentage.

In order to calculate a change, we have to analyze the data and our calculation step by step. First of all, we have to know if it is a decrease or an increase; then we have to tell the difference in percentage.

For example, the Dow Jones was 8,000 points yesterday and is 8,200 points today. The change is 200 points and we know it is an increase. Since it is an increase from yesterday's figure, we should compare it with yesterday's figure for the change in percentage.

- i) The change is 200 points (8,200 – 8,000)
- ii) The base is 8,000 points (we compare with yesterday)
- iii) The change in percentage is, therefore:

$$\frac{200}{8000} \times 100\% = 2.5\%$$

It is important to know what you are comparing and which data you should use as the base (denominator) when calculating the percentage.

First of all, you should know you are comparing the data at different times. Normally, it will be one in the past and one in the present time; or one in an older time and one a more recent time.

Secondly, you have to tell if the more recent data is larger than or smaller than the older one. If it is a larger one, there is an increase in percentage. If the more recent number is a smaller one, there is a decrease in percentage.

If you are not familiar with negative numbers (such as 10 – 3 is – 7), you may want to find out the difference by always minus the smaller number from the larger number. For example, if the Dow Jones was 6,000 points yesterday and is 5,700 points today. The change is 300 points (6,000 – 5,700) but you have to know that it is a <u>decrease</u>.

Once you know the difference, you can divide the difference by the older data. In the above example, the difference is 300 and the older data is 6,000; so the change in percentage is

$$\frac{300}{6000} \times 100\% = 5\% \text{ (decrease)}$$

Exercise X

a)   The oil price was $100 per barrel yesterday and is now $104 per barrel today. What is the change in percentage?

b)   The average price of houses was $200,000 in 2010 and it is now $210,000. What is the percentage increase?

c)   The average salary of a fresh graduate in 1988 was $20,000 per year and was $25,000 in 2004. What is the change in percentage?

d)   The average mark of a pervious test was 80 and the average mark of a current test is 76. What is the change in percentage?

e)   The temperature in the morning was 15 degrees and was 12 degrees in the afternoon. What is the change in percentage?

f)   The original price of a T-shirt was $50 and the sale price was $30. What is the change in percentage?

g)   A man's weight was 200 pounds and is now 160 pounds. What is the change in percentage?

Answers to Exercise X

a) The increase is $4; the change in percentage is 4%.

b) The increase is $10,000; the percentage increase is 5%.

c) The increase in salary was $5,000; the change in percentage is 25%.

d) The decrease in the mark is 4; the change in percentage is –5% (a decrease).

e) The decrease in temperature was 3 degrees; the change in percentage was –20% (a drop in temperature).

f) The reduction in price was $20; the change in percentage was –40% ('Less 40%' as a discount).

g) The loss in weight was 40 pounds; the change in percentage was –20% (a loss).

Converting Units

We use units of measurement every day – pound, kilogram, inch, meter, hour, year, dollar, cent, square foot, litre and gallon; just to name a few. It is our nature to use these units without fearing of making any mistake if no calculation is needed, such as to convert the unit from imperial to metric. All you have to do is to measure them.

Converting a unit from imperial to metric, for example, can also be a challenge for some people. How many pounds are there in 20 kilograms? How long is a 1 meter in feet? What is the temperature in Celsius when it is 72 degrees Fahrenheit? Once you know the relation between the units, you should be able to convert them without any difficulty. Here is how:

First of all, you have to know the basic relationships among the units:

$$1 \text{ inch } = 2.54 \text{ cm}$$
$$1 \text{ meter } = 3.28 \text{ feet}$$
$$1 \text{ mile } = 1.61 \text{ kilometre}$$
$$1 \text{ pound} = 454 \text{ gram}$$
$$1 \text{ degree Celsius } = 1.8 \text{ degrees Fahrenheit}$$
(the freezing point is 0 degree Celsius, 32 degree Fahrenheit)

Once you have the relationships, you can convert the units by using the simple trick in calculating ratio pair.
For example, you are going to convert 3.4 metres into feet. Let the answer be A feet and we use the ratio 1 metre to 3.28 feet, we have the equation:

$$1 \text{ metre to } 3.28 \text{ feet} = 3.4 \text{ metres to A feet}$$

Note: The units have to be in the same order in each side of the equal sign.

The trick we use in calculating ratio pair is *'the product of the outer pair equals the product of the inner pair'*.

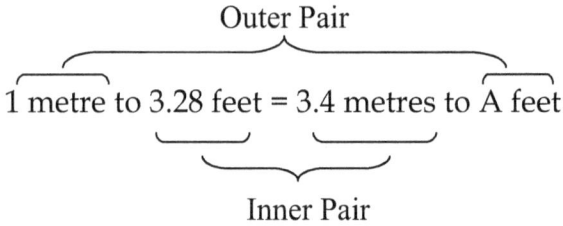

Outer Pair

1 metre to 3.28 feet = 3.4 metres to A feet

Inner Pair

That is,

$$1 \times A = 3.28 \times 3.4$$
$$A = 11.152$$

Therefore, we know 3.4 metres is 11.152 feet.

We can use the symbol ':' to denote ratio. 1 to 3.28 is denoted as 1 : 3.28 and 3.4 to A is 3.4 : A. The above equation becomes

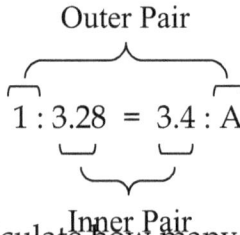

Outer Pair

$$1 : 3.28 = 3.4 : A$$

Inner Pair

Now we can try to calculate how many inches there are in 100 cm. Let the answer be B inches, we have

$$1 \text{ inch} : 2.54 \text{ cm} = B \text{ inches} : 100 \text{ cm}$$
$$1 \times 100 = 2.54 \times B \text{ (outer pair = inner pair)}$$
$$B = 39.37$$

That is, 100 cm = 39.37 inches.

Using the ratio pairs to convert units is an easy task. All you have to do is write down the ratio equation by using the same order of ratio and then use the '*the product of the outer pair equals the product of the inner pair*' rule to find out the unknown.

Converting temperature from degree Celsius to degree Fahrenheit, or vice versa is a little bit tricky. Most of the units start at the same point – zero. When we have nothing; the length is 0 cm, which is the same as 0 inch; the weight is 0 kilogram, which is the same as 0 pound; and etc. However, when we have 0 degree Celsius, it is not the same as 0 degree Fahrenheit.

Freezing point is 0 degree in Celsius scale, but it is 32 degree Fahrenheit. Therefore we have to use the equations below in converting them:

Celsius × 1.8 + 32 = Fahrenheit
(from Celsius to Fahrenheit)

(Fahrenheit – 32) ÷ 1.8 = Celsius
(from Fahrenheit to Celsius)

Examples:

a) What is the temperature in Fahrenheit when it is 30 degrees Celsius?

By using the first equation above, we have

Celsius × 1.8 + 32 = Fahrenheit

$$30 \text{ (C)} \times 1.8 + 32 = 86 \text{ (F)}$$

That is, 30 degrees Celsius is 86 degrees Fahrenheit.

b)   What is the temperature in Celsius when it is –16 degrees Fahrenheit?

By using the first equation above, we have

$$(\text{Fahrenheit} - 32) \div 1.8 = \text{Celsius}$$

$$(-16 - 31) \div 1.8 = -54 \div 1.8$$
$$= -30$$

That is, –16 degrees Fahrenheit is –30 degrees Celsius.

Exercise XI

a)   How long is 48 centimeters in inches?

b)   How long is 90 kilometers in miles?

c)   How long is 254 centimeters in feet?

d)   How long is 6 feet in centimeters?

e)   How much is 2,000 grams in pounds?

f)   How much is 7.2 pounds in grams?

g)   How much is 1 ounce in grams?

h)   How much is 15 grams in ounces?

i)   What is the temperature in Celsius when it is 500 degree Fahrenheit?

j)   What is the temperature in Fahrenheit when it is 100 degree Celsius?

k)   What is the temperature in Celsius when it is –40 degree Fahrenheit?

Answers to Exercise XI

a) 18.9 inches (1: 2.54 = X : 48)

b) 55.9 miles (1: 1.61 = X : 90)

c) 8.33 feet (1: 2.54 = X : 254; we get 100 inches)

d) 183 cm (6 feet is 72 inches, 1: 2.54 = 72 : X)

We can also use the meter to feet conversion
1 : 3.28 = Y : 6.
Y = 1.83 (meter), which is 183 cm.

e) 4.4 pounds (1 : 454 = X : 2,000)

f) 3,268.8 grams (1 : 454 = 7.2 : X)

g) 28.38 grams (as 1 pound = 16 ounces,
so 1 : 454 = 1/16 : X)

h) 0.53 ounce (as 1 pound = 16 ounces = 454 grams,
we have 16 : 454 = X : 15)

We can also use 1 : 454 = X : 15. X = 0.033 (pound),
which is 0.53 ounce.

i) 260 Celsius. Use (500 – 32) ÷ 1.8 = 260.

j) 212 Fahrenheit. 100 × 1.8 + 32 = 212

k) –40 Celsius. (–40 – 32) ÷ 1.8 = –40; this is the only
temperature that Celsius is the same as Fahrenheit.

The Rate

When we convert a unit to another unit, it is still the work on the same measurement unit – length, weight, volume or time. However, when we have to study the relationship between two units, it is a more difficult task as it involves changes in one unit over another unit of measurement.

For example, when we want to know how fast a car is moving, we talk about its speed and the unit is km per hour (or mile per hour); such as 90 km/h or 55 MPH. It is easy to read the speedometer and tell your speed, but it may be a challenge to find out the average speed by calculating the distance you travelled and the time you spent.

Similarly, when we want to tell how much money we are making, we will tell it by $15 per hour, or $600 per week, or $2,500 per month, or $30,000 per year. You should note that there are two units involved already – $ is the dollar sign, a unit for the money and the other one is the unit for time. Although the four pay rates are in different time units, basically they are the same (making $15 per hour is the same as making $600 per week, or $2,500 per month, or $30,000 per year).

Many people, without a good understanding of mathematics, find it hard to solve daily problems or academic questions involving rates. They do not know whether they should multiply or divide the numbers. In the case of division, they do not know which number to be divided by the other number (which one should be numerator and which one should be denominator).

For example, you are traveling at a speed of 120 km per hour, how long it will take to go to the destination that is 240 km away from your current position?

Assuming you need Y hours to go there; for every hour you drive, you will travel 120 km. Therefore, the total distance you will travel in Y hours is Y × 120. We know the distance is 240 km, so we have

$$Y \times 120 = 240$$
$$Y = 2$$

That is, it takes you 2 hours to arrive at the destination.

However, many people may not be able to figure out how to calculate the above question. Should it be 120 ÷ 240? Should it be 240 ÷ 120, or even 120 × 240? Some people just do not know when to use multiplication and when to use division. Assuming you do not know how to solve the question above, there is a simple way to help you by knowing the units of the objects involved.

First of all, we have to know the units of the objects we are dealing with and write them down. In the example above, assuming we do not know how to do the math. We know the speed is 120 km per hour, so the unit of speed is 'km/hour'. The distance from your current position to the destination is 240 km; the unit of distance is 'km'. The question asked how long it would take to go to the destination, so the answer would be Y 'hour'. In the above calculations, we missed the units in Y, 120 and 240; which is a bad habit.

That is, we should write down:

$$120 \ \underline{km/hour}$$
$$240 \ \underline{km}$$
$$Y \ \underline{hour}$$

When we compare the units, we know

$$km/hour = km \div hr$$

Since we have 120 km/hour, 240 km and Y hour and we should have the units of the equation as km/hour = km ÷ hr; therefore, the equation should be:

$$120 \ km/hour = 240 \ km \div Y \ hour$$
$$120 = 240 \div Y$$
$$Y = 240 \div 120$$
$$Y = 2 \ (hours)$$

Solving a problem by just comparing the units of the objects is not recommended, but it is the fastest way to solve a simple problem like that. Moreover, comparing the units of the objects is a good way to double check your answer to see if your calculation is correct. If the unit of your answer is not the one that the question asks for, you should know that your answer is wrong.

For example, you paid 100 dollars for 2 boxes of chocolates and each box contains 20 pieces of chocolates. How much is the price of chocolate per piece?

Assuming you are trying to solve this question and you let Y be the price of chocolate per piece. You then write the equation:

$$Y = 2 \times 40 \div 100$$
$$Y = 0.8$$

Now we can check if your calculation is right or wrong by checking the unit of Y. Since we want to know the price of chocolate per piece, it unit of Y should be $ per piece.

Now the equation above is actually

$$Y = 2 \text{ boxes} \times 40 \text{ piece/box} \div \$100$$
$$Y = 0.8 \text{ piece/}\$$$

You know that is wrong, as the unit is not the $/piece.

Exercise XII (Answers are on the next page)

a)   You walked 30 km in 5 hours. What is your average speed?

b)   You worked for 40 hours and got $600. What is your hourly wage?

c)   A train is travelling at a speed of 120 km per hour. What is the time required to travel a distance of 300 km?

d)   An aeroplane is flying at a speed 800 km per hour for 4 hours. How far has the aeroplane travelled?

e)   You paid $10 for 4 pounds of grapes. How much is the grape selling per pound?

f)   You spent 30 minutes to run from home to work, the distance was 5 km away. What is your average speed?

g)   A soft drink has two different sizes. The 2 litres bottle size is selling at $3 each and the 350 ml can size is selling at $3 for 6 cans. Which size will cost you less?

Answers to Exercise XII

a)   6 km per hour (30 km ÷ 5 hours).

b) $15 per hour ($600 ÷ 40 hours).

c)   2.5 hours (300 km ÷ 120 km per hour).

d)   3,200 km (800 km per hour × 4 hours).

e) $2.50 per pound ($10 ÷ 4 pounds).

f)   10 km per hour (5 km ÷ 0.5 hours).

g)   The price of 2 litres bottle size is $1.50 per litre ($3 ÷ 2,000 ml). The price of can size is $1.43 per litre [$3 ÷ (6 × 350 ml/1000)]. Therefore, the price of can size is lower.

Short Cut Summary

➢ Problem solving in daily life can be done by simply comparing the units of the objects.

➢ Comparing the units of the objects is a good way to double check your answer to see if your calculation is correct.

➢ For example, you would like to find out the speed (km per hour), all you have to do is to find out the units in the question. Since the unit is km/hour, that means we have to divide the number of kilometers travelled by the number of hours spent, then you will be able to find the answer. Common units to be used:

- o Length: cm, inch, foot, meter, km, mile
- o Weight: ounce, pound, gram, kilogram
- o Volume: cc (cubic centimeter), ml (milliliter), liter, gallon
- o Time: second, minutes, hour, day, week, month, year
- o Speed: km/hour (km per hour), mile/hour (MPH)
- o Interest rate: %/month or %/year (percent per month or year)

Put It in Writing

Not just solving complicated or difficult mathematics questions need to write them down; most of the time, you should write down the question you have and solve it step by step regardless how easy the question you may think.

Even mathematicians may not be able to memorize all the steps in a simple calculation, writing them down can solve the question faster and easier.

Writing down the details is the way to solve a problem logically, step by step. It is just like working on a puzzle, you need to group the pieces together and find out a way to construct the whole picture.

Algebra

Algebra is a subject that scares many people.   In fact, it is simply the concept of representing numbers by variables in high school level. Let us consider the following question:

After 12 years, John's age will be 4 times of his present age now. How old is John?

In order to solve this problem, we have to assume that John's age is X year old (note: we have to use an alphabet to represent the unknown in order to write the equation). Then, we have

$$X + 12 \text{ years} = 4 \text{ times of } X$$

That is,   X + 12 = 4X
                    ⇒   3X = 12;
                    ⇒   X = 4

Therefore, John is 4 years old.

This is a simple algebra question. By representing the age of John by the letter X, we have X + 12 is the age of John after 12 years. Since the age after 12 years will be the 4 times of John's present age; we have the equation X + 12 = 4X. We can then solve the equation and find out that X = 4.

Writing the problem down is the first step in elementary algebra. You have to let a variable (X or Y or any alphabet) to represent the unknown, form an equation using the information given by the question, and using that equation to solve for the unknown.

The steps for simple cases are:

1. Let X be the unknown.
2. Express the relationship of X with other data given by the question to form an equation.
3. Solve the equation to get the unknown.

Examples:

a)   I have some candies. If I get 20 more candies, I will have 5 times the candies that I have now. How many candies do I have?

        Let X be the number of candies I have. We can then represent the sentence in the question by an equation:

$$X + 20 = 5X$$

$$\Rightarrow \quad 4X = 20$$

$$X = 5$$

Therefore, I have 5 candies.

b) It takes me 12 hours to drive from home to the airport at normal speed. If I drive 20% faster, how soon I can get to the airport?

The steps for questions more than one unknown can be:

1. Let a, b, c, … be the unknowns (the number of unknowns depends on the question).
2. Express the relationships of the unknowns to form some equations according to the question.
3. Solve the equations to get the unknowns.

Let a be my normal speed; and
Let b be the time I need when driving 20% faster.

Distance from home to airport = 12xa (normal speed)
= bx1.2a (20% faster)

$$\Rightarrow \quad 12a = 1.2ab \quad \text{(the distance is the same)}$$

$$\Rightarrow \quad b = 10 \text{ (the time needed is 10 hours)}$$

Exercise XIII (Answers are on the next page)

a) A house is sold for a price which is $400,000 more than the purchase price 20 years ago and the new price is 500% of the old one. What is the original purchase price?

b) A house is sold for a price which is $100,000 more than the purchase price 10 years ago and the new price is 300% of the old one. What is the new sale price?

c) You drank 300 ml of water, which was 75% of the bottle. How much water did the bottle hold at first?

d) A toy was sold for $272 at a discount of 20%. What was the list price?

e) The total number of hens and cows on a farm is 72. The total number of legs of them is 208. How many hens and how many cows are there in the farm?

f) Two men had a race. X was running at 10 km/hr while Y at 8 km/hr. Y was allowed to run 1 hour ahead of X. How long did it take X to overtake with Y?

g) If the interest on a loan for 5 years at a rate of 10% per annum was $3,000, how much was the loan amount?

Answers to Exercise XIII

a) $100,000

Let X be the original price. X + $400,000 = 5X.
X = $100,000.

b) $150,000

Let X be the original price. X + $100,000 = 3X.
X = $50,000. New price is $50,000 + $100,000

c) 400 ml

Let Y be the amount of water the bottle held at first.
Y × 75% = 300 ml. Y = 400 ml.

d) $340

Let Y be the list price. Y × 80% = $272. Y = $340.

e) There are 40 hens and 32 cows

Let X be the number of hens and Y be the number of cows.

X + Y = 72 (total number of them is 72)
2X + 4Y = 208 (hen has 2 legs and cow has 4 legs)

From the first equation, we have X = 72 – Y and we substitute this into the second one.

$2(72 - Y) + 4Y = 208$
$144 - 2Y + 4Y = 208$
$2Y = 64$
$Y = 32$, therefore $X = 40$

f)   4 hours

Let the time that X could takeover Y be a hours (from the time Y started).

Since X started 1 hour before Y, the distance X had run after a hours after Y started was 8 km/hr × (a + 1) hrs.

The distance Y had run after a hours was 10 km/hr × a hrs.

When Y could takeover X, it was the time Y met X. That is, the distances they had run were the same. Therefore $8 \times (a + 1) = 10 \times a$

$8\,a + 8 = 10\,a$
$8 = 2\,a$
$a = 4$

g) $6,000

Let Y be the loan amount.

Y × 10% per year × 5 years = $3,000
Y = $6,000

The Triangle

Although you may write down your own equation to solve a problem, sometimes it is necessary to memorize a formula in order to write your equation. For example, you have to know the formula of calculating the area of a circle, a rectangle and a triangle; or the formula of calculating simple interest, the rate of return and etc.

Using a triangle to represent such formulas is an easy way to memorize them; adult students find it particularly useful as the triangle helps them visualize the formulas.

For example, the area of a rectangle is its length multiply by its width.

$$\text{Area} = \text{Length} \times \text{Width}$$

To denote this formula, we can use the triangle below:

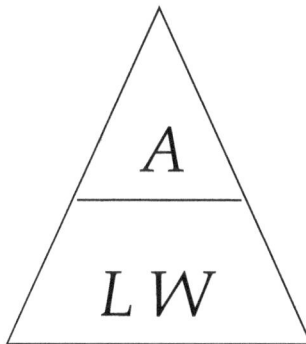

$$\frac{A}{L\,W}$$

The line in the middle of the triangle separating the A and LW actually represent the fraction line. When we want to know how to get W, just cover W in the triangle and we get

$$\frac{A}{L} \quad ; \text{that is } W = \frac{A}{L}$$

Similarly, when we cover L, we get

$$\frac{A}{W} \quad ; \text{that is } L = \frac{A}{W}$$

when we cover A, we get

$$L\,W \quad ; \text{that is } A = L\,W\,(L \times W)$$

Therefore, if we know any two of the data: area, length and width; it is easy to find the unknown one by using the triangle.

Another example is to calculate the volume of a block, the formula is Volume = Length × Width × Height.

To denote this formula, we can use the triangle below:

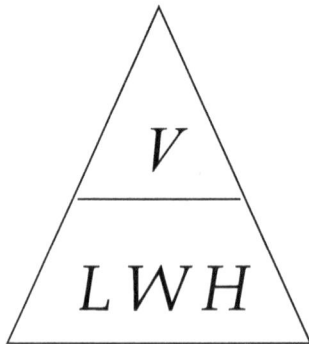

We have: $V = L\,W\,H$; or

$$\frac{V}{LH} = W \quad ; \text{ or}$$

$$\frac{V}{LW} = H \quad ; \textit{or}$$

$$\frac{V}{WH} = L$$

Where V is Volume, L is Length, W is Width and H is Height.

Basically, you can fit any formula involving simple multiplication or division into the triangle.

Examples:

a)   Interest = Principal × Interest Rate

$$(I = PR)$$

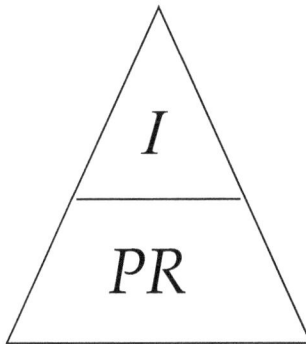

b)   Gross Profit Margin Ratio = Gross Profit ÷ Sales

Gross Profit Margin Ratio × Sales = Gross Profit

GP = Margin × Sales

or simply

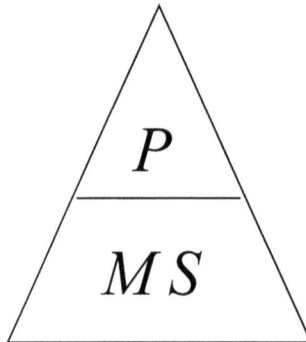

Where P is Gross Profit, M is Margin and S is Sales.

Exercise XIV (Answers are on the next page)

Draw a triangle to represent the formula below:

a)   The circumference of a circle is 2 times its radius and π.

b)   Area of a circle is π times the square of its radius.

c)   The rate of return of an investment is the income divided by the amount of investment.

For example, if you have $4,000 in a saving account and the interest on it is $500 per year. The rate of return of this saving account is 12.5% ($500 ÷ $4,000)

d)   The time (number of years) to double your investment is approximately 72 divided by the rate of return of your investment (this is called the *Rule of 72*).

For examples, if the rate of return of your mutual fund is 6%, it takes 12 years to double your money (72 ÷ 6 = 12); if the rate of return of your mutual fund is 10%, it takes 7.2 years to double your money (72 ÷ 10 = 7.2).

Answers to Exercise XIV

a)

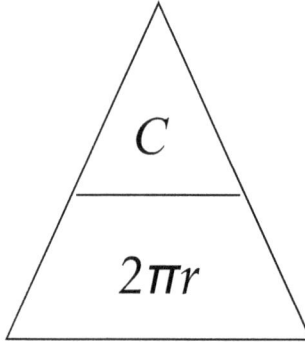

Where C is a circumference, r is the radius.

b)

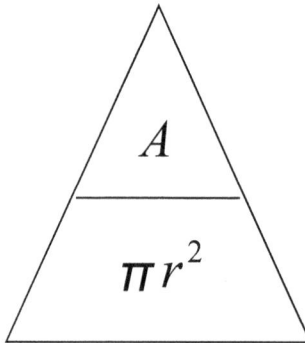

Where A is an area, r is the radius.

c)

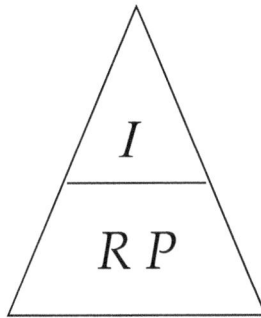

Where *I* is the income, *R* is the rate and *P* is the price of investment (invested money).

d)

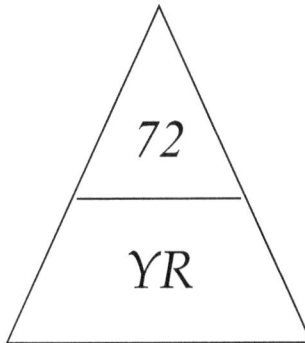

Where 72 is in percentage, Y is the number of years to double your investment and R is the rate of return of your investment (also in percentage). In general, the formula is:

$$Y = \frac{72}{R}$$

Short Cut Summary

➢ Problem solving can be done by simply putting the problems in writing with the unit in mind:

- o Speed: km/hr or mile/hr (km per hour or mile per hour)
- o Salary: $/month or $/year (dollars per month or year)

➢ The steps for solving questions with the unknown can be:

- o Let a, b, c, ... be the unknowns (the number of unknowns depends on the question).
- o Express the relationships of the unknowns to form some equations according to the question.
- o Solve the equations to get the unknowns.

➢ You can use a triangle to help you memorize the formula. For example, Area = Length × Width, you can draw a triangle like this one:

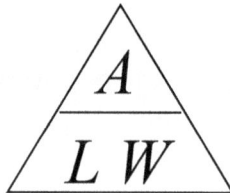

$$\frac{A}{L \; W}$$

Where A is Area, L is Length and W is Width. Cover the unknown and you will get the formula for it.

Simple Logics

In layman terms, logic is the study of reasoning. It helps people decide whether a statement is true or false. Having a good logical sense will bring you success in not only mathematics, but also in any field that requires arguing, critical thinking, or negotiation. Careers such as lawyers, journalists, accountants and even medical doctors need logical minds to be successful.

Simple Statements

A simple statement is a sentence with a value, either true or false. For example, *'all humans are warm blooded creatures'* is a statement and we know it is true. The statement *'humans are cold blooded creatures'* is false.

'At least one man is warm blooded' is true. Although all men are warm blooded, *'at least one man is warm blooded'* will not make it a false statement as ALL men are warm blooded of course means AT LEAST one man is warm blooded.

'A creature is not warm blood is not a human' is true because warm blooded is, in fact, one of the criteria to define human beings. This is why we have the first statement *'all humans are warm blooded creatures'*. Therefore, if a creature is not warm-blooded, that creature cannot be a human.

However, if a creature is warm blooded, that creature may be or may not be a human. Animals such as lions, dogs and birds are all warm blooded. This kind of argument is the basic of logic.

Simple Arguments

When there is a statement, such as the above one:

>   *All humans are warm blooded creatures.*

or

>   *Human implies Warm Blooded*

or

>   *If you are a human, you will be warm blooded.*

Its reverse statement is:

>   *All not warm blooded creatures are not humans.*

or

>   *Not Warm Blooded implies Non-human.*

or

>   *If you are not warm-blooded, you will be a non-human.*

All reverse statements are actually the same argument as the original statement. If you can prove either one of the statements is false, then both statements are false. Of course, if one statement is true, both statements are true.

>   If A implies B, then not B implies not A.

Example:

Consider the below statement:

Your IQ will be over 140 if you are a university graduate.

Its reverse statement is:

You are not a university graduate if your IQ is lower than 140.

To prove that the first statement is false, either we can find a person among all university graduates whose IQ is below 140 or to find a person with IQ below 140 who is a university graduate.

Since the first statement is false, the second statement is false as well.

Backward Argument

When we have an argument: If A, then B. The backward argument is: If B, then A.

Example:

If it is a rainy day, then it is not a sunny day.

The backward argument is: *If it is not a sunny day, then it is a rainy day.*

From the example above we know that even if an argument is true; the backward of it may not be true. For the example above, if it is not a sunny day; it can be a snowy day, a cloudy day, but not necessarily a rainy day. Therefore, the argument is not true.

For the first statement, we use: *All humans are warm blooded creatures.* Its backward argument is

> *All warm blooded creatures are humans.*

 or

> *Warm Blooded implies Human*

Clearly, there are many warm blooded creatures in the world and the argument above cannot be true.

Sometimes a backward argument may be true. Consider this statement:

A colourless and tasteless liquid with 0 degree freezing point, 100 degrees boiling point, the density of 1 gram per cubic centimeter implies that the liquid is water.

The backward statement of it is:

A liquid is water when it is a colourless and tasteless liquid with 0 degree freezing point, 100 degrees boiling point, the density of 1 gram per cubic centimeter implies.

When a statement (argument) is true and its backward is also true, then this kind of implication is called <u>if and only if statement</u>.

If and Only If

We deal with possibilities everyday. Who took my newspaper? Who was the last one to leave office yesterday? Who will get promoted? Which elevator will come first?

There may be more than one possibility in each of those questions. In other cases, there may be only one possibility. For example, a couple is living in an apartment by themselves; if the wife cannot find the book she just left on the table, it must the husband who took it.

The wife could say:

If the book was not on the table, it was because my husband took it.

This kind of statement has only one cause and only one possible result, so the backward is also true:

If my husband took the book, it would not be on the table.

The wife can use one sentence to conclude the case:

The book was not on the table if and only if my husband took it.

This kind of deduction is very important, especially for some professions such as lawyers, medical doctors and computer programmers. For example, pneumonia patients may have common symptoms such as high fever, chills, chest pain and cough. However, a patient who has all these symptoms may not be a pneumonia patient. More symptoms are needed to conclude whether the illness is pneumonia or not.

Doctors use simple logic to deduce what the illness is; lawyers use it to make their arguments, and computer programmers must use it to teach computers how to 'think'.

Mathematical Notations

It will be simpler to write a logic statement by using mathematical notations. For example, the statement *You are warm blooded if you are a human* can be expressed as

You are a human $\Rightarrow$ You are warm blooded

The notation $\Rightarrow$ means 'implies'. The statement reads as

You are a human implies you are warm blooded.
For an *if and only if* statement such as *The magazine was not on the table if and only if my husband took it* can be expressed as

The magazine was not on the table $\Leftrightarrow$ *my husband took it*

The notation $\Leftrightarrow$ is a sign that means 'implies and being implied by'.

'And' and 'Or'

In our daily life, we always use 'and' and 'or' to describe people and things. When we say "Boss asks you and me to go to that meeting"; that means both you <u>and</u> me have to go to that meeting. When we say "Boss says the meeting needs you or me"; that means only one of us has to be in the meeting, but not both.

In fact, 'and' and 'or' are operators to link statements together which make logical statements have different meanings. Many people get it right when it is an ordinary statement, but get confused when it is a logical statement (a condition).

For example, when we say "Don't drink and drive"; you know it is okay to drink, OR to drive, but not both. When you don't want people to bring any food and any drink to your store, should you say "No Food and Drink" or "No Food or Drink"?

It is interesting that these two statements appear the same to most of us as we know the meaning of them – no food and no drink. However, if you apply the logic of "No Drink and Drive", then you should know the first statement simply means 'no food and drink at the same time'. "No Food or Drink" means either food or drink is not allowed. A correct statement using the operator 'and' is "No Food and No Drink".

Example 1:

A crowd of people are classified into 3 categories (a person can belong to more than one group) – Group A: Below the age of 21, Group B: 18 to 65 years old and Group C: over 60 years old.

a) Discount coupons are delivered to people that belong to Group A or Group C. That means people in the age of 0 to 21 and over 60 will be eligible for the discount.

b) A lucky draw will be given to people that belong to both Group B and Group C. That means only people from 61 to 65 will be eligible for the draw.

c) No water will be given to people that belong to all three groups (A and B and C). That means no one will be penalized by that – no one belongs to all three groups.

Example 2:

'And' and 'or' are important in ordinary calculations too, the following equation and inequalities may give you the ideas when to use 'and' and when to use 'or'.

a) $(X - 2)(X - 4) = 0$

We need either $(X - 2) = 0$ or $(X - 4) = 0$

$\Rightarrow$ $(X - 2) = 0$ or $(X - 4) = 0$

$\Rightarrow$ $X = 2$ or $X = 4$

b) $(X - 2)(X - 4) > 0$ (greater than 0)

We need both $(X - 2)$ and $(X - 4)$ to be positive or both to be negative.

$\Rightarrow$ $(X - 2) > 0$ and $(X - 4) > 0$; or
$(X - 2) < 0$ and $(X - 4) < 0$

$\Rightarrow$ $X > 2$ and $X > 4$; or
$X < 2$ and $X < 4$

$\Rightarrow$ X > 4 or X < 2

c)   (X – 2) (X – 4) < 0   (smaller than 0)

We need one positive and one negative to make the product a negative one.

$\Rightarrow$ (X – 2) > 0 and (X – 4) < 0 ; or
      (X – 2) < 0 and (X – 4) > 0

$\Rightarrow$ X > 2 and X < 4; or
      X < 2 and X > 4 (impossible)

$\Rightarrow$ 2 < X < 4 (X is greater than 2 but smaller than 4)

Exercise XV
(Answers are on the page after these questions)

If the first statement of the following is true, will the second statement be also true?

a)   When my tummy is having a noise, it implies that I am hungry.
When I am hungry, my tummy will make noise.

b)   When my teacher gives me candies, it means she likes me.
That teacher will give me candy if she likes me

c)   All green vegetables are good for health.
Poison vegetables are not green in colour.

d)   Whenever there is a ball, the baby will play with it.
The baby does not play; it means there is no ball.

e)   Every man has a beard.
Leslie has a beard, Leslie is a man.

f)   If you are Japanese, you must love sushi.
If you are not Japanese, you may still love sushi.

g)   If a rat falls into water, it will drown.
If a creature falls into the water and does not drown, it is not a rat.

h)   A piece of paper is water proof if and only if it is coated.

A piece of paper is not water proof; that means it is not coated.

Give the value(s) of X if

i)   X > 2 and X > 5

j)   X > 2 or X > 5

k)   X > 2 and X = 3

l)   X > 2 or X = 3

m)   X > 2 and X ≥ 2

n)   X > 2 and X ≤ 2

o)   X > 2 or X ≤ 2

p)   X > 2 or X ≥ 2

q)   X < 2 and X > 5

r)   X < 2 or X > 5

Answers to Exercise XV

a)   Not true. The first statement means: My tummy will not have any noise if I am not hungry.

b)   Not true. The first statement means: If my teacher does not like me, she will not give me candies.

c)   True.

d)   True.

e)   Not true. The first statement means: Whoever has no beard is not a man.

f)   True. All Japanese love sushi does not mean that no one else can love sushi.

g)   True. If it is a rat, it will drown.

h)   True. The other statement will also be true: A piece of paper is not coated; that means it is not water proof.

i) $X > 5$

j) $X > 2$

k) $X = 3$

l) $X > 2$

m) $X > 2$

n)   No solution (X cannot be both smaller than 2 and greater than or equal to 2)

o)   X can be any number (real number)

p) $X \geq 2$

q)   No solution (X cannot be both smaller than 2 and greater than 5)

r)   X can be any number but not between 2 and 5 (except in the range $2 \leq X \leq 5$)

Short Cut Summary

➤ Logic is the study for reasoning. It is the foundation for an argument.

➤ Logic Rule #1:

If A implies B, then Not B implies Not A

- ○ If A $\Rightarrow$ B, then Not B $\Rightarrow$ Not A
- ○ E.g.   All humans are warm blooded.
        Not warm blooded $\Rightarrow$ Not human

➤ Logic Rule #2:

If and only if statement: both the statement and its backward statement are true.

- ○ If A $\Rightarrow$ B and B $\Rightarrow$ A, then A $\Leftrightarrow$ B.
- ○ That means only 1 possibility – Not A $\Leftrightarrow$ Not B.

➤ AND means both statements (conditions) have to be met.

➤ OR means either one statement (condition) has to be met.

Part II – Games

Multiples of Ten – Solitaire Game

Speed plays a very important role in the calculation. You may know how to solve a simple calculation when it is asked in front of a group, but your answer will be 'meaningless' if someone else has already told it loud.

Since we use decimal as the numerical system, it is crucial to computing 10 with addition and subtraction. Therefore, we should get familiar with the number 10 as soon as we can; say, under the age of 6.

Multiples of Ten is a funny game, plus it can tell your luck. You will use a deck of playing cards, only the cards from ace to 9 (36 cards) will be used to play.

You flip the cards out one by one, and to add up all the points (ace is 1). When the sum is a multiple of 10, the cards should be collected and placed as a small deck. You then repeat the process again and when the sum is a multiple of 10, the cards are collected and placed as another small deck.

When all 36 cards are played, it should make a perfect match so that no more cards will be left on the table. Otherwise, you should have made a wrong calculation.

If you have more than 4 decks of finished cards, it will mean that you have a good luck. ☺

Fishing Game

Fishing Game is an excellent game to train small kids to sum up a pair of numbers to 10; it works with older kids as well as adults too. The game is very simple and you use a deck of playing cards, treating ace as 1.

It is a game for 2 to 5 players. When there are only 2 players, 10 cards are assigned to each player; when there are 3 players, 7 cards are assigned to each player; when there are 4 players, 5 cards are assigned to each player; only 4 cards will be assigned to each player if there are 5 players.

Twelve cards should be put on the table with face up, regardless of the number of players (except the case of 3 players, only 10 cards should be put on the table). The remaining cards are left on the table to be picked by the players.

All players take turns by matching a card on the table with the player's card on hand to make a 10. That is; 1 will match with 9, 2 will match with 8, 3 will match with 7, 4 will match with 6, 5 will match with 5. For ordinary playing cards, 10 will match with 10, Jack will match with Jack, Queen will match with Queen and King will match with King.

There are two chances for each player in each turn. The first chance comes from the player's own card. When a player can match a 10 with one card on hand and one card on the table, the player will take both cards. If the player cannot make a 10, the player has to leave a card on the table from the player's hand.

The same player will draw a card from the deck of the remaining cards. If that card can match with a card on the table to form a 10, then the player can take both cards. If not, that card has to be left on the table.

After all the players have played their cards on hand, the remaining cards should also be used up. Players can count how many points they get from the cards (the total of the numbers, J, Q and K count as 10). However, only the red cards have points, black ones are zero.

Hint: To get higher marks, players should target red cards instead of a black one.

Adding Up Game

Adding Up is an excellent game to play fast and get a result. A very simple game but requires a higher level of adding skills than Fishing Game. Again, you use a deck of playing cards, treating ace as 1. Only the cards from ace to 10 (40 cards) will be used to play. The cards King, Queen and Jack (12 cards) will be used as 'vouchers' to award to players and will be placed on the desk as the Bank.

It is a game for 2 players or more, up to 13 players. Each player will get 3 cards and they have to add the three cards together and only the last digit will be counted as a score. For example, the sum of the cards 6, 8, 9 is 23 and will be treated as 3. The score is therefore 3.

After all, players have counted their points, the one(s) with the highest score wins (there may be more than one winner if their points are the same). The winner will get 1 mark voucher – Jack is worth 1 mark, Queen is worth 2 marks and King is worth 4 marks. Winners can take the voucher from the Bank and can make the change if necessary.

Once all the vouchers have been awarded to winners, players can tell who the winner is by counting the total marks won.

Four Minus One Game

Four Minus One is a game for 2 players or more, up to 10 players. It is the extended version of Adding Up, the rules are the same except that each player will get 4 cards and only 3 cards can be used to add up the points.

For example, if you get 3, 5, 6 and 9; then you should use 3, 6 and 9 to add up the points. The total point is 18 and your score is 8. If you get 5, 6, 7 and 8; you should use 5, 6 and 8 to add up and you will get 19; your score will be 9.

All the other rules are the same as in Adding Up game.

Top Up Game

Top Up is a game for 2 players or more, as long as all of them can see the cards on the table.

A deck of playing cards, only the cards from ace to 10 (40 cards) will be used to play. Again, ace will be treated as 1 and the cards King, Queen and Jack (12 cards) will be used as 'vouchers' to award to players.

One of the players will be responsible for delivering the cards. Three cards should be drawn from the deck and flipped at the same time (not one by one) so that all the players will see all three cards at the same time.

The task is to top up the sum of the three cards to the nearest ten's multiple, and the first person gets the correct answer wins the game.

If the three cards are 1, 6, 9; then the number to top up will be 4 (to make it 20). If the three cards are 5, 8, 10; then the number to top up will be 7 (to make it 30). If the three cards are 3, 7, 10; then the number to top up will be 0 (the sum is 20, a multiple of 10 already).

This simple game involves both addition and subtraction and is a funny game that you can enjoy and train your skills in a short period of time.

The Ghost Legs

Ghost Legs is a matching game designed to pair up things. It is often used to distribute things among people, to assign partners or to pick one person (or a few people) from a group to perform a task.

It is a diagram consisting of vertical lines with horizontal lines connecting two adjacent vertical lines scattered randomly along their length; the horizontal lines are called "legs". The number of vertical lines will equal to or more than the number of people playing and at the bottom of each line, there is an item - a thing that will be paired with a player. The bottom of some of the lines may be empty, that means the player who chose such a line will have empty hands. Figure 5 is an example of such.

The rule of the game is to choose a line on the top and follow this line downwards. When the vertical line encounters a horizontal line, you must follow it to get to another vertical line and continue downwards. Repeat this procedure until you reach the end of a vertical line. You are then given the thing written at the bottom of the line.

For example, ten people were going to decide their order in performing a task. They used the diagram (Figure 5) and one picked 'F' to start. The route would be the highlighted one, which went down from 'F' and would go to destination '1' (Figure 6). Other people chose other starting points would end up to other destinations. If you know the trick of it, you can always get to the destination you want.

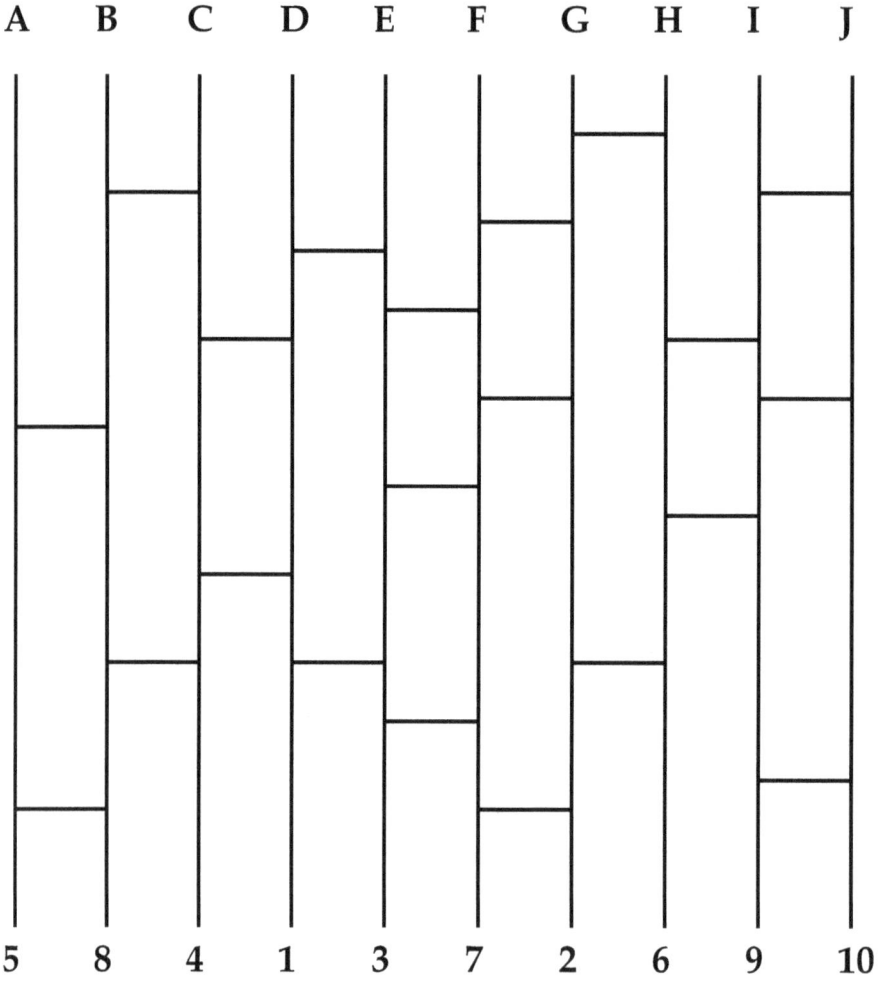

Figure 5

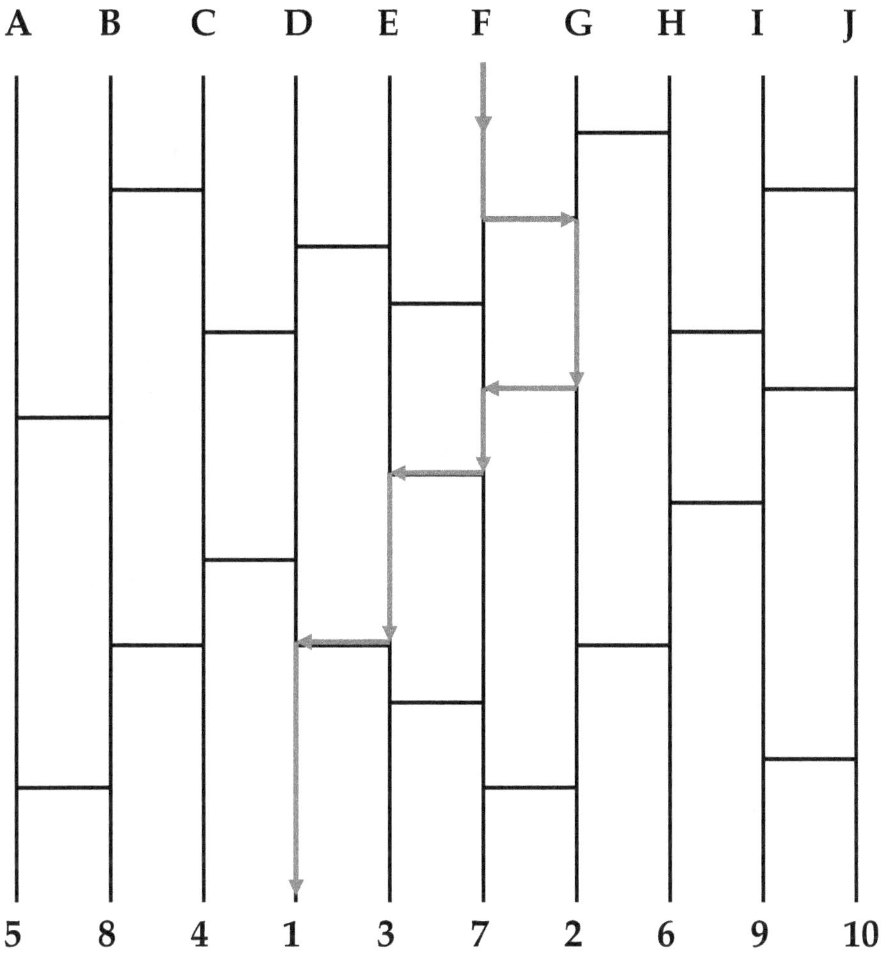

Figure 6

In fact, every starting point will lead to a unique ending point. Having known that, it is easy to `predict` the result by working backwards.

For example, if you want to get '1' as your destination; you can start backward from '1' and moving up, using the same rule. That is, to follow that line upwards. When that vertical line encounters a horizontal line, you must follow it to get to another vertical line and continue upwards. You will eventually go to 'F' as the destination (Figure 7).

You can, therefore, pick the preferred outcome in advance by using this backward trick. For example, if you want to have '3' by using the legs in Figure 5; you will know that you should pick 'H' to start (by going upwards from '3').

There is, however, one way to prevent people from using this trick – to place a piece of paper to cover the middle part, so that people cannot tell the routes by going backward. The reality is that not too many people know the trick and you can take advantage of knowing it.

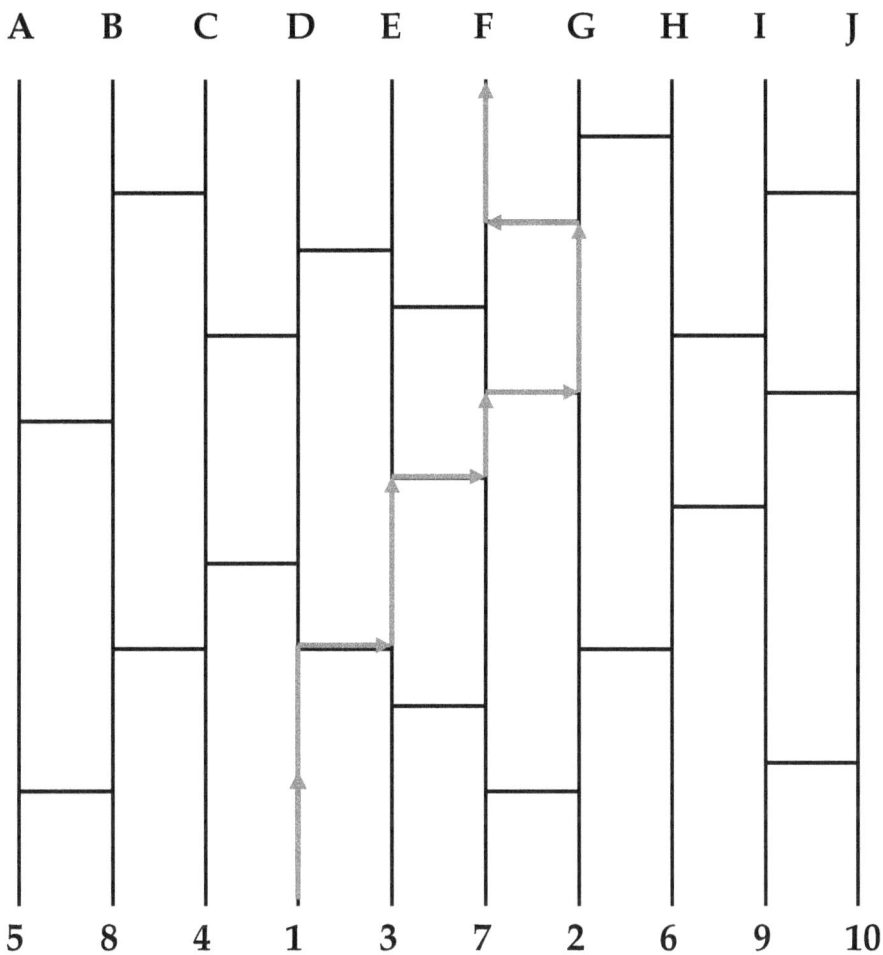

A B C D E F G H I J

5 8 4 1 3 7 2 6 9 10

Figure 7

Exercise XVI

Use the Ghost Legs on the next page to find:

a)   The ending point of A.

b)   The ending point of B.

c)   The ending point of C.

d)   The ending point of D.

e)   The starting point of 2.

f)   The starting point of 7.

g)   The starting point of 9.

h)   The starting point of 10.

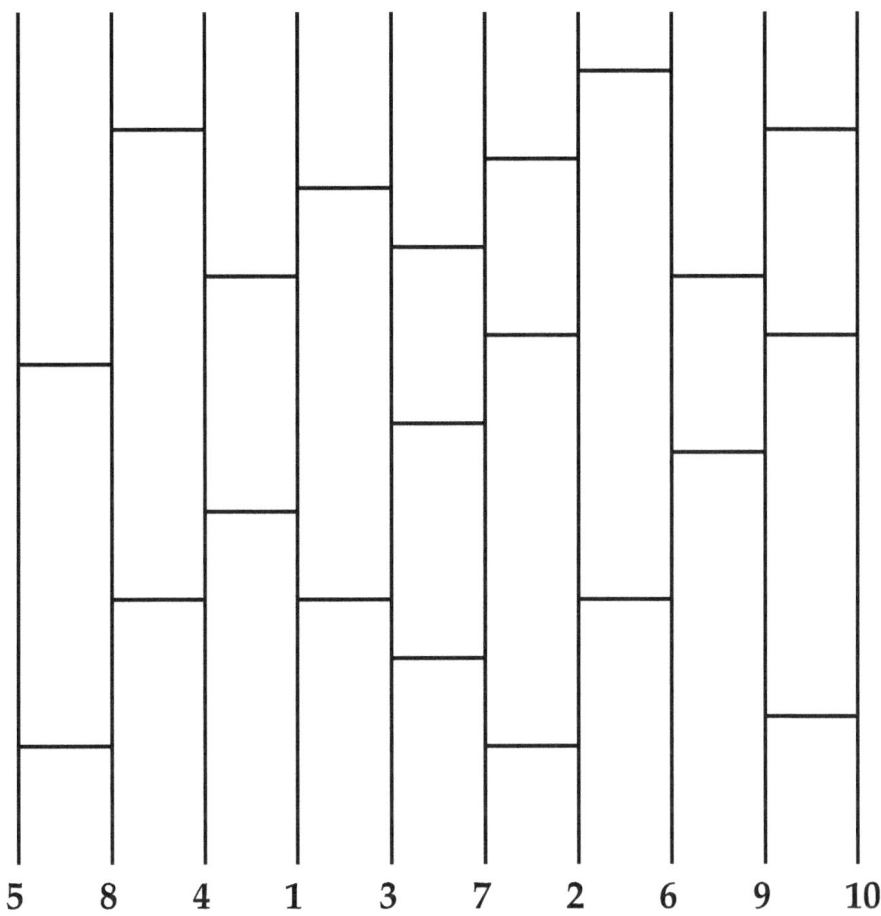

A   B   C   D   E   F   G   H   I   J

5   8   4   1   3   7   2   6   9   10

(Answers are on the next page)

Answers to Exercise XVI

a)   4

b)   5

c)   8

d)   6

e)   E

f)   I

g)   G

h)   J

Conclusion

Mathematics is not a difficult subject, at least not in the university or lower levels. However, many people find it boring and afraid of it. In fact, mathematics can be interesting. As it applies to our daily activities – school, work, leisure and more; it is necessary to have a good sense of math.

Arithmetic skills can be trained by repeating the exercises, or through interesting games. It takes time to memorize the important skills such as the multiplication table. It may be boring or time consuming, depends on your personality. However, once you can memorize the table, you will be more confident in many different types of calculations.

Like all other games, basketball, football and chess; mathematics needs practice. The games introduced in this book are the foundation of arithmetic and can train your mind to be sensible to figures. Parents are encouraged to play those games with their children until they are familiar with all the basics.

To conclude, mathematics is not a difficult subject; it just needs you to get familiar with. Practice is the key to success.

www.ingramcontent.com/pod-product-compliance
Lightning Source LLC
Chambersburg PA
CBHW060043210326
41520CB00009B/1245